A Basic Training, Caring & Understanding Library

Feline Behavior

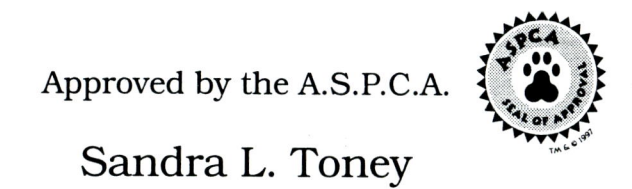

Approved by the A.S.P.C.A.

Sandra L. Toney

Published in association with T.F.H. Publications, Inc.,
the world's largest and most respected publisher of pet literature

Chelsea House Publishers
Philadelphia

A Basic Training, Caring & Understanding Library

Kitten and Cat Care
Healthy Skin & Coat: Cats
The Myth & Magic of Cats
Persian Cats
Feline Behavior
Obedience Training
The Well-Trained Dog
Adopting a Dog
The Perfect Retriever
Healthy Skin & Coat: Dogs
Traveling With Dogs
Training Older Dogs
Housebreaking & Other Puppy Problems
Puppy Care and Training
Your Healthy Puppy
Perfect Children's Dogs
Training Your Puppy
You & Your Puppy

Publisher's Note: All of the photographs in this book have been coated with FOTOGLAZE™ finish, a special lamination that imparts a new dimension of colorful gloss to the photographs.

Reinforced Library Binding & Super-Highest Quality Boards

This edition © 1998 Chelsea House Publishers, a division of Main Line Book Company.

© yearBOOKS, Inc.

135798642
Library of Congress Cataloging-in-Publication Data applied for 0-7910-4810-1

LIBRARY
HOLBEIN SCH.
MT. HOLLY, NJ

FELINE BEHAVIOR

Sandra L. Toney

The Publisher wishes to acknowledge the following owners of cats in this book: Gloria and Susan Adler; Ann Marie Andriaur; Janet Bassetti; Kim Brantley; Kaye Chambers; Andrew DePrisco; Susan and Ken Eklund; Gail Fine; Brenda Goessling; Meredith Gowell; Babette Gray; Geri Hamilton; Margery S. Hoff; Linda B. Jones; Oliver H. Jones; Louis Johnhson; Christine W. Keightley; Marilyn R. Knopp; Kelly Mayo; Melissa Mattison; Carol Rothfeld; Caroline Scott; Betsy Stowe; Frank A. Szablowski; Sandra Toney; and Karen West.

Photographers: Isabelle Francais; Michael Gilroy; Gilian Lisle; Robert Pearcy; Betsy Stowe; and Sandra Toney.

What are Quarterlies?

Books, the usual way information of this sort is transmitted, can be too slow. Sometimes by the time a book is written and published, the material contained therein is a year or two old...and no new material has been added during that time. Only a book in a magazine form can bring breaking stories and current information. A magazine is streamlined in production, so we have adopted certain magazine publishing techniques in the creation of this Cat Quarterly. Magazines also can be much cheaper than books because they are supported by advertising. To combine these assets into a great publication, we issued this Quarterly in both magazine and book format at different prices.

© by T.F.H. Publications, Inc.

Distributed in the UNITED STATES to the Pet Trade by T.F.H. Publications, Inc., One T.F.H. Plaza, Neptune City, NJ 07753; on the Internet at www.tfh.com; in CANADA Rolf C. Hagen Inc., 3225 Sartelon St. Laurent-Montreal Quebec H4R 1E8; Pet Trade by H & L Pet Supplies Inc., 27 Kingston Crescent, Kitchener, Ontario N2B 2T6; in ENGLAND by T.F.H. Publications, PO Box 15, Waterlooville PO7 6BQ; in AUSTRALIA AND THE SOUTH PACIFIC by T.F.H. (Australia), Pty. Ltd., Box 149, Brookvale 2100 N.S.W., Australia; in NEW ZEALAND by Brooklands Aquarium Ltd. 5 McGiven Drive, New Plymouth, RD1 New Zealand; in SOUTH AFRICA, Rolf C. Hagen S.A. (PTY.) LTD. P.O. Box 201199, Durban North 4016, South Africa; in Japan by T.F.H. Publications, Japan—Jiro Tsuda, 10-12-3 Ohjidai, Sakura, Chiba 285, Japan. Published by T.F.H. Publications, Inc.

MANUFACTURED IN THE
UNITED STATES OF AMERICA
BY T.F.H. PUBLICATIONS, INC.

Quarterly

yearBOOKS, INC.
Dr. Herbert R. Axelrod,
Founder & Chairman

Barry Duke
Chief Operating Officer

Neal Pronek
Managing Editor

yearBOOKS are all photo composed, color separated and designed on Scitex equipment in Neptune, N.J. with the following staff:

DIGITAL PRE-PRESS
Patricia Northrup
Supervisor

Robert Onyrscuk
Jose Reyes

COMPUTER ART
Patti Escabi
Sandra Taylor Gale
Candida Moreira
Joanne Muzyka
Francine Shulman

ADVERTISING SALES
Nancy S. Rivadeneira
Advertising Sales Director
Cheryl J. Blyth
Advertising Account Manager
Amy Manning
Advertising Director
Sandra E. Cutillo
Advertising Coordinator

©yearBOOKS, Inc.
1 TFH Plaza
Neptune, N.J. 07753
Completely manufactured in
Neptune, N.J. USA

Designed by Sandra Taylor Gale
Cover design by Sherise Buhagiar

Contents

Your Aging Cat 5

Early Learning and
Socialization Period ... 12

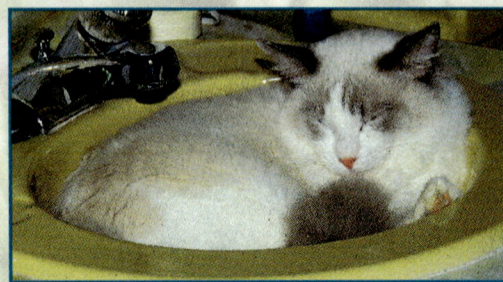

Introduction to Cat Behavior ... 3

Gender and Genetics 18

Breed and Behavior ... 23

Common Kitty Quirks
and How to Handle
Them 37

Cats at Play 53

Suggested Reading 64

INTRODUCTION TO CAT BEHAVIOR

As the renowned 16th century writer Cervantes once declared, "Those who'll play with cats must expect to be scratched." There is no doubt that this insightful author must have owned a lot of felines in his life to have made such an accurate observation.

Cats have never been—nor will they ever be—easy to understand. Why they exhibit certain behaviors has mystified feline experts and ordinary kitty owners alike through the ages. Many factors contribute to the emergence of a cat's distinct personality, and in ensuring that no two cats will ever be the same.

Age, gender, breed, and parentage play crucial roles in determining a cat's personality, as do such factors as how early the cat was taught certain tasks and when it initially socialized with humans. As devoted cat owners, we can delve into each of these factors individually in an attempt to understand our beloved companions, and to offer explanations for their captivating behavior. Examining the diverse, complicated mind of the creature known as the cat offers us a lifetime of loving appreciation for this unique feline who has touched our lives.

Throughout the ages, cats have fascinated their human companions with their beauty and unique personalities.

4 YOUR AGING CAT

Establishing "house rules" during kittenhood will help to ensure a successful training program.

YOUR AGING CAT

One of the most important factors influencing your cat's behavior throughout his life is his age. Each rung on the ladder of life plays an integral role in developing Kitty's personality and behavior. From a kitten's first few weeks of life—and well into his advanced years—you can expect certain specific behaviors during each stage of your cat's life.

NEWBORNS
When a kitten is born, he is completely helpless and relies on his mother for every want and need. Since kittens are born blind and practically deaf, they actually depend on their mother for their very survival during the first few weeks of life. Eating and sleeping are a kitten's sole functions, and their mother provides nourishment through nursing, as well as a safe and peaceful shelter in which to sleep. Of course, this helplessness soon passes as the kitten discovers that running, jumping and playing are much more fun than just hanging out with Mom.

KITTENHOOD
There probably isn't a cuter sight to behold than a fluffy, adorable, two month-old kitten frolicking through your house.

Your cat can live with you for a very long time. Be certain to take good care of him and give him the very best veterinary care possible.

YOUR AGING CAT

Playing is more than just having fun, it is an important part of a kitten's growth and development.

However, this mischievous side to your precious baby will not always emerge when it is most convenient for you. When you are fast asleep at 2:00 a.m., and anticipate a big day at work the following morning, your cat will not always cooperate with your schedule.

Kittens are oblivious to timing and responsibility. If your kitten wants to jump on your head and claw your nose at 2:00 a.m., he will do it, regardless of how you feel about it. You won't even be consulted.

Owning a kitten is not a job to be taken lightly. Kittens are rambunctious and have unlimited energy when it comes to uncovering for the first time the joys the world has to offer. Everything is brand new to them and your home provides an abundance of adventures for the curious kitten. What an amazing revelation for the cat to discover that the roll of toilet paper will go around and around and is perfect for shredding!

The perfect age to acquire a kitten is after he has been weaned from his mother and can eat solid food on his own—generally after eight weeks of age.

Taking a kitten away from his mother and siblings too early can result in an emotionally "handicapped" animal who may not have been taught such "natural" behaviors as hunting and playing.

It is a proven fact that outside cats live much shorter lives because of the inherent dangers of the outside world. Since your kitten will most likely spend most of his days within the safe confines of your home or apartment, he will need plenty of toys and other mental stimuli to develop his instinctive hunting skills, which never really go away, regardless of whether kitty is hunting a real mouse or a toy one.

Although your kitten will

Your new kitten will be frightened when you first bring him home and will need time to become acclimated to his new environment.

YOUR AGING CAT

need an assortment of toys to play with (the more the better so he won't get bored) you should discourage him from playing with your hands or feet. When a kitten is a tiny, delicate creature, it may be cute to see him attacking your hand—which is sometimes bigger than himself. But, as your kitten ages, his claws and teeth become much bigger and sharper, and one day you'll realize it is no longer cute when he attacks your hand—it hurts!

If you let him use your hand as a toy when he is young, he will not understand why he can't use it as he matures. Cats are creatures of habit; it is not wise to make human hand-attacking a habit because it will eventually be painful. So, don't encourage this aggressive behavior with your kitten, and you will not have to stop it later on.

The most important point to make about your kitten is that you've acquired him in the most impressionable and trainable stage, and, unfortunately, the most fleeting stage as well. Enjoy your little kitten because it doesn't take long for him to reach middle age!

ADULTHOOD

If you are currently thinking about getting a new cat for your household, you might want to consider acquiring an adult feline. Most people automatically consider bringing a little kitten into their home, but, as mentioned earlier, you must be willing to invest a lot of time in training.

The decision to adopt an adult cat, however, means not having to expend nearly as much time and effort in teaching and acclimating him to your home. All you have to do is spend a few short hours with an adult cat to determine his distinct personality and disposition.

Of course, the onset of adulthood signals your cat's sexual maturity. Between six and twelve months of age, a cat will reach sexual maturity and, in the case of the female, pregnancy is almost an absolute. Even if you think you can confine your female, a cat in heat will be a most annoying animal, and she is likely to make every possible attempt to escape.

The only solution to this problem is to have your female cat spayed, which will result in a much more agreeable pet to have around the house. And, considering the millions of unwanted cats who land in shelters and alone on the streets, spaying is a logical solution.

A whole male cat, also known as a "tom cat," will become aggressive and unruly at the onset of sexual maturity, and he will also attempt to escape

When choosing a kitten, check for the following: bright clear eyes with no signs of discharge, a dry nose, and clean ears with no signs of foreign matter.

YOUR AGING CAT

Adult cats make excellent, loving companions and need time to play. If you do not have the time to be around, why not have two cats that can keep each other company?

your home at every available opportunity. This signals the beginning of a behavior known as "spraying," wherein the unneutered male backs up against an object and urinates on it in a territorial behavior used to establish his domain. The smell from male cat spraying is offensive and pungent, and most people cannot live with a male who sprays in their home. Avoid this behavior by having your male cat neutered as early as your veterinarian deems possible. Once your cat begins this marking behavior, neutering may not stop him from continuing this habit for the rest of his life.

Regardless of whether you've owned your cat since kittenhood, or have just recently acquired him, once your cat has reached adulthood you will have an excellent and loving companion who still needs time to play and, naturally, to sleep. These are good years for your feline. Savor them because, sooner than you think, Kitty will no longer have that bounce in his step and, although he may not be eligible for social security, he will definitely be considered a senior citizen.

RETIREMENT YEARS

Considering the advances in veterinary medicine, as well as the exceptional care provided by today's cat owners, the life expectancy of indoor felines is now as high as 15 to 20 years. A feline ages most rapidly during the first two years of life, with maturation becoming more moderate until the later years. During the elderly stage of your cat's life, he will need special love and attention. Getting older is not easy for any creature, and it is certainly not easy for your cat to endure the mental and physical decline of aging.

YOUR AGING CAT

Your cat will defend himself and his territory from other animals in a number of ways. This feline's puffing up is just one of the ways he wants to tell visitors that he is the boss!

YOUR AGING CAT

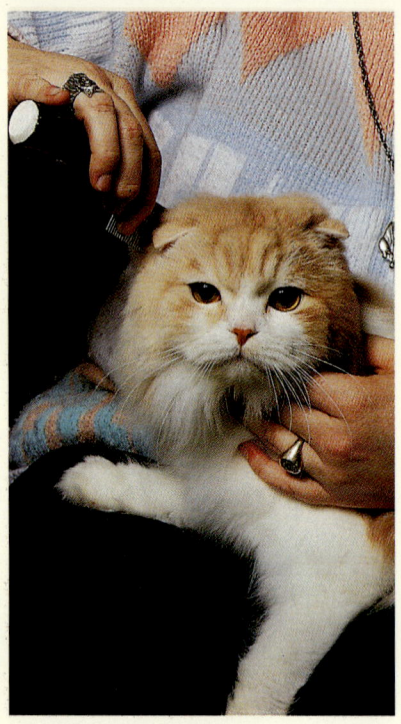

As your cat ages, the most noticeable change will be his activity level. As an adult, your cat will have a greater need to sleep than he did as a kitten.

But you can make your cat's golden years a prime time of life if you anticipate the changes that may occur in this final stage of his long and happy life. Physical deterioration becomes more and more pronounced as the aging cat experiences diminished capacity in such areas as sight, hearing and muscular function. These bodily changes will most likely result simultaneously with behavioral changes as the cat senses that he's slowly losing control of many aspects of his life.

An aging cat's energy level is probably the most noticeable change in a feline's behavior, because the metabolic rate declines at a slow but steady rate over the years. And, although it seems almost impossible to believe that an elderly cat could sleep any more hours in a day than a younger cat (who sleeps about 16 hours a day when in top form), he will feel the need to sleep even more hours per day. He will tire very easily and may need your assistance in jumping onto familiar places like your bed or the sofa that were easily reached in the past.

It should be pointed out that animals age at a different rate than humans. So, a cat who has lived to 12 years of age (in human years) is approximately 64 years old in cat years. As all feline lovers know, cats have a tremendous amount of pride. For the geriatric cat, the inability to jump and play as he once did is probably the hardest thing for him to accept. So, if he misses a jump onto the sofa and falls flat on his behind, it's best to pretend that you didn't even notice the mishap. He'll probably pick himself up off the floor and try to convince you that he actually *meant* to miss that jump while practicing his falling maneuvers.

The most you can do for your senior feline is give him even more love and respect than you did in his younger years, and allow him to enjoy his final days in comfort and peace. After all, it's the least you can do to thank him for all the years of loyalty and companionship he's given you.

Owning a cat since kittenhood and watching him grow up forms a mutual bond of respect and love between owner and kitty.

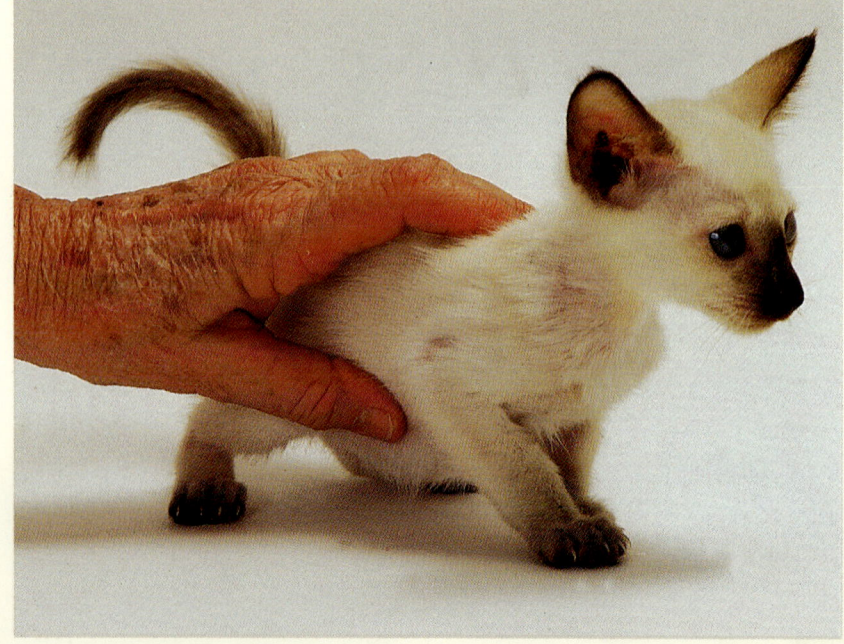

YOUR AGING CAT

The most you can do for your senior feline is give him even more love and respect than you did in his younger years, and allow him to enjoy his final days in comfort and peace.

EARLY LEARNING AND SOCIALIZATION PERIOD

Besides considering age as a determinant in a feline's personality, researchers have also studied the role that early learning and socialization play in a kitten's development. In other words, does it make a difference when a kitten is introduced to humans? And, does it matter if a kitten is introduced to one person or a multitude of people? And finally, are there certain periods of time when a kitten is more inclined to learn certain things? Surprisingly enough, the answer to all of these questions is yes.

EARLY HANDLING

Wouldn't it be a tough job to go to work everyday and have to "play" with kittens? Believe it or not, there are people who actually perform this task in scientific laboratories—day in and day out—in an effort to pinpoint the exact age that a kitten should be introduced to humans in order to maximize their sociability. Intense and controlled studies, such as those pioneered by Dr. Eileen Karsh at Philadelphia's Temple University, have determined that holding or handling a kitten between the ages of two and seven weeks makes for the most sociable adult cat. Kittens handled before or after this interval were deemed less sociable in their interactions with humans later in life.

Dr. Karsh's study noted that many of the kittens who were initially handled after the two to seven week period ultimately behaved like kittens who had not been handled at all. Although it seems unlikely that such a small difference in time could produce such significantly varied behavior, the evidence is overwhelming.

Dr. Karsh's study also found that the amount of time the kittens were handled played a pivotal role in overall sociability. The obvious conclusion was that the longer a kitten was held per session, the friendlier the kitten behaved towards people over time. And, it was discovered that it does not really matter who handles the kitten because kittens do not necessarily become "attached" to one particular person, per se, but to humans in general. As long as the kitten is handled by someone

Kittens that are introduced and socialized with humans at an early age are more sociable than those that are not.

EARLY LEARNING AND SOCIALIZATION

during the critical socialization period (i.e., the second through the seventh week), the kitten is likely to react favorably to most any human's touch, because he will have formed an attitude of general acceptance towards humans.

This crucial socialization period is not just important in interacting with humans; it pertains to other animals as well. For example, kittens learn a lot from interacting and socializing with their littermates, and those who have been orphaned or are the sole offspring in the litter have been found to be much more insecure and fearful of other felines as adults.

Kittens who spend their socialization period with siblings—or even non-related kittens for that matter—turn out to be more sociable towards other cats as they mature. The same is true of kittens who come in contact with dogs and other animals. These cats will almost always accept another animal as long as it has been introduced during the specified socialization period.

MULTIPLE HANDLING

Most kitten owners want their pets to be friendly and even-tempered towards them and to visitors in the home as well. Obviously, when company arrives, it is far preferable to call your cat into the room to show off instead of watching him cower fearfully underneath the furniture. The key to acclimating your cat to people in general (and not just you or people they often see) is, once again, related to the critical two to seven week socialization period.

Kittens learn a lot not only by interacting with humans but from interacting with their littermates as well.

Dr. Karsh's study at Temple University reveals interesting information regarding the handling of these kittens during this critical socialization period. First of all, her study found that the number of handlers holding the kitten influenced the kittens' behavior, such that if a variety of people held a kitten for the required amount of time each day—as opposed to the same person—the kitten was far more likely to approach unrecognized humans, and did so sooner than kittens who had been handled by

If your kitten has parents who are shy, chances are he will be shy too. Likewise, if the parents are outgoing and friendly, you can expect the kitten to be the same.

EARLY LEARNING AND SOCIALIZATION PERIOD

The Sphynx is a hairless breed of cat that has a super personality. Remember to expose your kitten to a number of different experiences so that he will become a well-adjusted adult.

the same person each day. Additionally, a kitten who had been held by the same person each day would allow that person to hold him for much longer periods of time than someone who had not handled the kitten each day.

Learning to understand your cat's personality, his moods, and instincts is one of the most interesting aspects of cat ownership.

But Dr. Karsh's study was not the first of its kind. Over thirty years ago, scientists conducted similar experiments with Siamese kittens and found that those who had been handled during the socialization period seemed to mature faster than those who were not. Not only did they leave their mother's nest earlier than those who had not been handled, but they were not as fearful of unrecognized people and objects, and approached the unknown far more assertively.

Plus, although all Siamese kittens are born with completely white coats, it was noted that those kittens who were handled during the early socialization period acquired the distinct Siamese colors and point markings more rapidly as well. So, it is clear that the ideal setting in which to initiate kittens to the socialization process

Make certain your cat is friendly and non-aggressive toward all people. Hissing and growling with mouth wide open, an angry cat can appear a formidable foe to its aggressor.

occurs within the second to seventh weeks of life, and involves a variety of people handling them for a significant amount of time.

It is also important to introduce these youngsters to a variety of different experiences (such as car rides, rambunctious children, vacuum cleaners, lawn mowers and other noisy social situations) so they can grow up to be well-adjusted, good-natured adults who can adapt to the diverse circumstances of everyday life.

EARLY LEARNING

Every grade school science student becomes

EARLY LEARNING AND SOCIALIZATION

familiar with the experiments conducted by Pavlov, the scientist who "conditioned" dogs to recognize food by the sound of a bell. He trained these dogs by literally hitting a bell before feeding time, so that eventually, the mere sound of the bell prompted the dogs to salivate in anticipation of a meal.

Cats are just as capable of learning. Although dogs will perform tasks purely for the intrinsic reward in securing their owner's love, cats, on the other hand, seem to learn new tricks faster and more readily if given food as a reward. Just as they need to be handled at a certain stage of kittenhood, they also learn faster and more readily during that time frame as well.

Many people talk about cats and their "instincts," when, in actuality, this is a result of behaviors that are learned very early in life. A kitten's most important teacher is his mother and, by watching her, he learns the basics of how to become a cat.

One of the most meaningful lessons a mother cat teaches her kittens is the art of successful hunting. Since cats have only been domesticated some 5,000 short years, they still prepare themselves for life in the wild, which in modern times means survival without can

Mutual grooming is one behavior that can be observed frequently in households that have more than one cat.

openers and the cessation of cat food dispensed at the first "Meow!"

A good mother cat will show her kittens how to catch, kill, and eventually eat their own food. This lesson starts very early—in fact, it occurs during the critical socialization period.

All cats, no matter the age, need an outlet for their energy. These two Tonkinese are definitely up to something.

Cats are the greatest imitators, and inevitably learn a task if they see it done repeatedly. (If you have never witnessed a mother cat bringing her kittens a mouse, you have missed the ultimate teacher enlightening her very "willing" students.)

For a time, the mother cat will bring back dead rodents and creatures for her kittens to smell, bat around, and perhaps even taste. However, the teaching phase soon ends as the mother starts bringing back live prey so she can show her kittens how to kill and eat the catch.

To any observer, this may seem like a cruel ritual, but in fact, it is quite necessary for these kittens to learn to hunt, kill, and eat their prey in case they should ever find themselves out in the real wild, or even exposed to the "urban jungle," as many

EARLY LEARNING AND SOCIALIZATION PERIOD

stray and homeless cats find themselves today.

Teaching the kitten to play is a significant learned behavior that helps the kitten adapt to many aspects of life. Playing, in essence, affords kittens the opportunity to hone their hunting skills. A kitten may first turn his attention to the mother's swaying tail, delightfully pouncing on it at every opportunity. Soon, however, a kitten's littermates become the favorite objects of entertainment, and such interaction is indeed crucial, especially in view of studies that have shown that kittens born without littermates and deprived of play often become socially "retarded."

So, it is quite obvious that the length of time handled, the variety of people interacting with the kitten, as well as what the kitten is able to see and learn during the critical second to seven weeks of life, cumulatively contribute to the cat's overall development. These factors influence their lifelong attachment to human beings and the manner in which our felines enjoy and conquer the world around them.

Playing is hard work! Kittens will spend hours with each other—playing and wearing each other out.

GENDER AND GENETICS

17

Maine Coon. The large eyes, high set cheekbones, and lightly feathered ears are the distinguishing facial characteristics of Maine Coons.

GENDER AND GENETICS

Discovering how behavior develops in any creature is a task not to be taken lightly. Myriad details are involved in deciphering your cat's personality, and three of the most critical elements are genetics, gender and the component of alteration. Parents naturally determine whether a feline is male or female, but whether the kitten has been altered or is still whole also plays a pivotal role in resultant behavior. Research into each of these topics is ongoing, and will always invite debate regarding how significantly each factor impacts your cat's overall behavior.

Gender, genetics and the component of alteration all play a role in your cat's personality and his behavior.

GENETICALLY SPEAKING

Whether man or beast, every being is composed of a certain gene pool unique to the individual. Of course, the kitten's appearance is determined by genes passed on by his parents, but is it possible that actual behavior can be linked to the mother and father as well? Many feline behavior experts believe the answer to that question is yes. It is difficult to determine just how much of a kitten's behavior is genetically linked, and how much is determined by the early learning factor, wherein a kitten learns from his mother and mimics her behavior. However, numerous studies have shown that mother cats do indeed transfer behavior to their kittens.

It is apparent that the mother's attitude plays a vital role in determining her kittens' behavior towards humans. Dr. Eileen Karsh's study revealed that if a mother cat shies away from people, her kittens will also shy away. Conversely, if the mother acts calmly and non-apprehensively towards people, her kittens will emulate this comfort level in their interaction with humans as well.

Whether this is a learned behavior or an inbred trait remains a mystery because studies have lacked conclusive evidence regarding the father's influence in the disposition of progeny.

Dr. Sandra McCune, an animal behavior expert at

It is difficult to determine how much of your cat's behavior is genetically linked, and how much is determined by the early learning factor.

GENDER AND GENETICS

the Waltham Centre for Pet Nutrition in the United Kingdom, has conducted studies designed to prove the existence of a genetic link between a particular litter and their father. In her study, she produced two different tom cats to sire two different litters of kittens. One of the toms was quite unfriendly towards people while the second was very friendly and people-oriented. The researchers found that, although none of the kittens had ever had any contact with their fathers, they exhibited the same type of temperament as the tom who had sired them. This would appear to be possible only if behavior had been genetically passed from one generation to the next.

Because of the feline over-population crisis, and the undesirable habits of unaltered felines, it is best that you do not keep your cat whole unless you are an experienced breeder.

ALTERATION ADVANTAGES

With millions of cats put to sleep every year because of the feline overpopulation crisis, there is virtually no reason to keep a cat whole, or unaltered, unless you are an experienced breeder. If you own a pedigree cat that you want to breed, make sure you have good homes for all the kittens after their birth.

Many people want kittens because they are cute and lovable, but they soon grow into adult cats who will only produce many more kittens in their lifetime if they remain unaltered. We, as responsible pet owners, have the ability to control the problem to some extent by taking our felines to the veterinarian and having them altered. It is a simple, relatively inexpensive procedure designed to curb overpopulation.

Conclusions vary in regard to behavioral changes in altered cats,

If you are the owner of pedigree cats that you want to breed, make sure you have homes for all of the kittens before you plan the breeding.

GENDER AND GENETICS

By neutering your male cat at a young age, you will reduce the possibility of his trying to escape to find females in heat, as well as avoid the bad habit of urine spraying before it begins.

but experts agree that you can expect to find certain common behaviors in altered cats. Owners often complain about "spraying" (wherein male felines produce a pungent-smelling urine used to mark their territory). All cats are territorial, but unaltered males are especially so in order to protect their area from other tom cats who threaten their domain or compete for a female in heat. The next cat to come along also marks the spot in a communication ritual that has been in existence since the origin of cats.

While it's true that the tom's territorial obsession will never completely disappear, it can be reduced considerably once the cat is neutered and the inclination to compete for females in heat is removed. The key factor to consider in this instance is the cat's age: If you wait too long to have your cat neutered and he's already begun to mark his territory, he may never stop. And it's not because he needs to compete with other males; it has simply become a habit.

Cats rely on their sense of smell to a much larger extent than humans, which explains the practice of territorial marking. So, if there happens to be a female in heat in your neighborhood, chances are that your indoor tom will catch the scent in the air and do everything in his power to attract the fertile female.

The solution therefore is to have your male cat neutered at a young age. Check with your veterinarian to determine the appropriate time for this procedure. After neutering, you'll also find that your cat is far less likely to attempt to escape, as he no longer seeks females in heat.

Anyone who has ever seen a bloody, scratched tom after a cat fight will also realize another reason to have a tom cat neutered: When a female is in heat (or estrus) male cats will travel for miles to meet her, which, of course leads to competition for her attention. Males will fight ferociously to become the victor and win the queen's favor. A female cat in heat becomes loud and is in a distressed state several times a year, and efforts to confine either a male or female in heat will probably prove futile, as mating and reproduction triggers a deep-seated instinct that defies restriction.

Given the opportunity to

As heartwarming as a new litter of kittens can be, the responsibility, expense and lack of demand all point to spaying your pet.

GENDER AND GENETICS

21

By spaying or neutering your cat, you not only will help to reduce the over-population crisis, but you will have a more enjoyable companion who wants to spend time with you.

escape, the queen will follow her instincts and flee in order to mate, and, in all likelihood, become pregnant. The result will be more cats in a world where there are not enough homes for the ones already in existence. Therefore, neutering not only controls the cat population but it also has a calming effect on the cat's behavior.

As with any area of science where studies must be conducted and results are based on observation, some experts believe that altering a cat will not alter its personality. However, many more experts agree that once you've stopped the cat's overwhelming urge to reproduce, your feline will be less stressed, less aggressive, and absolutely more loving and cooperative.

Once you neuter your male or spay your female, a new way of life begins, and your cat will be able to concentrate on his/her favorite person—you! And, doesn't every cat owner desire a comfortable, affectionate, agreeable cat who has nothing to prove other than his never-ending love for you.

GENDER CHARACTERISTICS

Once you have altered your feline and have witnessed behavioral changes in your cat, you may want to look a bit more closely at the gender of your beloved companion. Although some behaviorists believe that once the kitten has been spayed or neutered, he becomes virtually "genderless," there are still others who believe that definite personality differences exist between male and female felines, despite alteration.

One area of contention will always be how demonstrative a cat is—and to whom. Some cat fanciers believe that female felines are more affectionate and loving pets, and that they are more likely than males to be lap cats who require more attention and petting from those she knows well. On the other hand, another faction argues that male cats are far more affectionate and loving, and much more inclined to be sociable to people in general. It has even been suggested that "opposites attract," and that, in theory, female felines relate better to male humans, and vice-versa. The debate continues in both scientific and non-scientific circles, with a definite bias on the part of proud feline owners!

The roles played by gender, genetics, and alteration, and what they have to do with overall personality, are still being studied by behavior experts and everyday cat owners alike. While feline behaviorists continue to study the mysterious beast known as the cat, owners of this clever creature conduct their own informal studies into the psychology of their cat's sometimes odd, but always captivating, behavior.

Most animal behaviorists agree that once your cat has been spayed or neutered, you will have a more affectionate, loving, and more socially inclined pet.

BREED AND BEHAVIOR

The breed of cat you choose also plays an important part in determining the cat's overall personality. Different cats exhibit distinct personality traits peculiar to their breed. Of course, it is impossible to state that all Persians are one way and all Siamese another. No two cats will ever be exactly the same, but each breed has attributes that have lead to certain generalities in behavior. The following are common personality traits found in the most widely recognized breeds in the world.

ABYSSINIAN

Even if you have never been a feline fancier, owning an Abyssinian might just change your viewpoint. The Abyssinian, whose ruddy coloring is the most popular variety, is a very intelligent, outgoing cat who loves to interact with people. However, it is not a cat that should be left alone for long periods of time because, without activity and intellectual stimulation, he will easily become bored and restless.

This cat is also very affectionate to his favorite humans but doesn't like to have an abundance of other felines around to share the spotlight. If an Abyssinian spends long periods of time alone, he will fret and worry until the two of you are reunited. So, even someone who was once less-than-enthusiastic

The head of the American Shorthair should be broad with well-developed cheeks and eyes and ears set well apart.

BREED AND BEHAVIOR

The name American Curl comes from the bend of the cat's ears, which must be at least a 90-degree arc, but never greater than 180 degrees.

about cats in general will surely fall in love with a cat as loyal and devoted as the Abyssinian.

AMERICAN CURL

Unlike the Abyssinian who needs a great deal of human interaction, the American Curl does not insist on such constant attention, although it makes for a very good friend and companion. This breed tends to be friendly and smart, and given the unusual distinction of curled back ears, they also make interesting conversation pieces. The American Curl will not lack in energy, and likes to have regular periods of playtime. This feline is quite adaptable to most any situation and is known for his curiosity. The breed accepts other animals remarkably well, but, as with most pets, still needs its fair share of love and attention.

AMERICAN SHORTHAIR

This is a very popular cat because of its low maintenance reputation and striking beauty—especially the silver tabby featured in many commercials and advertisements. The American Shorthair is normally a quiet cat who doesn't require as much attention as other breeds. This feline is perfectly satisfied living in close quarters, like an apartment, but, because of his excellent hunting skills, will probably bring home quite a few rodent trophies.

The American Shorthair has been called the "Gentle Giant" because of his even temper and loving disposition. Although they can be outgoing, American Shorthairs are normally so quiet and agreeable that they are considered rather shy.

BENGAL

The Bengal is a most unusual looking cat because of its spotted leopard appearance that comes from crossing a domestic cat with an Asian Leopard cat. This breed of cat is highly intelligent and loves to hunt prey, whether he is stalking a catnip-filled mouse or chasing a string you dangle before him.

Bengals are also friendly, affectionate, and extremely curious, so make sure you provide your Bengal with plenty of mental and physical stimulation for him to be his absolute happiest!

BIRMAN

Known as a feline who is

This Bengal kitten exhibits a desirable coat pattern, an expressive face and a loving personality. Bred by Andrew DePrisco and Barbara J. Andrews, this five-week-old kitten is out of Topspot's Simba.

BREED AND BEHAVIOR 25

The Bombay was originated as a hybrid of the Burmese and the American Shorthair, but its distinct features have earned the breed a following all its own.

This British Shorthair shows the well-rounded head with round eyes of a brilliant gold color.

extremely tolerant of children and other pets, the white-pawed Birman is a medium-to-large-sized cat whose roots can be traced back to Burma, where it was considered a sacred temple cat. The Birman is a smart and clever creature, but is not as loud and demanding as such other breeds as the Siamese. The Birman has a patient and carefree nature, and is very tolerant of children.

BOMBAY

The Bombay was created by a breeder who wanted to possess her own little black panther. This sociable cat has been likened to a monkey because of its acrobatic tendencies and playful nature.

The only color accepted in the Bombay class is black, and so it very much resembles a little black panther. These felines are more easily trained than most cats, and will quickly learn to do such tasks as fetching, walking on a leash, and jealously guarding its territory. The Bombay also makes a great lap cat.

BRITISH SHORTHAIR

The British Shorthair is very much like its American counterpart, the American Shorthair. An easy-going, sweet animal, the British Shorthair enjoys the company of humans but does not feel the need to "talk" to them frequently.

This breed of cat is very reserved and quiet, and is happy either surrounded by people or left alone. The British Shorthair is a rather "modest" cat, and doesn't require the constant pampering and fussing of certain other breeds.

BURMESE

Dog lovers will surely adore a Burmese feline because they have many of the qualities admired in canines. They simply love people to a greater degree than most cats. Large, innocent-looking eyes can mesmerize owners into granting this feline's every wish. But you will be generously rewarded in return as your Burmese will want to be by your side every second of the day. Whether helping you write a letter or keeping you warm under the covers at night, if you want a devoted dog, but in a cat's body, the Burmese is for you!

The head of the Burmese should be rounded without any flat planes. The face should look full with the round eyes, the more brilliant the eyes, the better.

BREED AND BEHAVIOR

The Chartreux is a French breed that is agile and refined and can be very affectionate.

CHARTREUX

If a Burmese isn't available for the dog lover in you, then the Chartreux may be a suitable alternative. This cat, which is commonly a muscular, solid adult, is a friendly, devoted feline, and behaves somewhat like a dog when rough-housing with the kids or other household pets. Yet, the Chartreux is a very devoted pet and will remain loyal to you and your family throughout his lifetime. The small, delicate voice that comes from this massive animal will surprise you as you will expect nothing less than a lion's roar from such a robust cat. Although both males and females make excellent pets, Chartreux breeders normally agree that the altered male is the preferred choice among pet owners.

CORNISH REX

Most cat lovers will agree that there is no such thing as an ugly cat, although this is sometimes the reaction to a Cornish Rex. With huge ears on a small head and a curly fur coat so short that the breed almost appears hairless,

The body of a Devon Rex is long, slender and medium sized, with a very hard muscular frame.

the cat has even been referred to as "alien-looking."

Once you get to know the Cornish Rex, however, you will come to believe that there really is no such thing as an ugly cat. The affectionate, giving personality of this creature makes him lovable inside and out. The Cornish Rex keeps his kittenish ways through adulthood, as he continues to race around the house and play any game you are willing to teach. To the Cornish Rex owner, he is truly the most beautiful of all breeds.

DEVON REX

The Devon Rex is usually described as having a "pixie-like" appearance. This cat has an amicable disposition and, like the Cornish Rex, at first glance, has sometimes been considered rather homely because of its unusual appearance.

Once again though, when you get to know a Devon Rex, you will begin to appreciate his unique personality. He is very protective of his owners, and almost always remains

The Egyptian Mau is the only naturally spotted domestic breed. This silver Mau shows the contrast in coat color for which the breed is known.

BREED AND BEHAVIOR

The tail of the Japanese Bobtail should be upright when the cat is relaxed, and the hair on the tail should be slightly longer than the hair on the rest of the body. The head should have a triangular shape with gentle curves.

nearby wherever you are in the house. Unfortunately however, both Cornish and Devon Rexes have been known to be on the "naughty" side occasionally. One look into those saucer-sized eyes though, and you are sure to melt into forgiveness.

EGYPTIAN MAU

Since the Egyptian word for "cat" is "mau," it follows that this fascinating feline, once glorified in ancient Egyptian hieroglyphics, is appropriately called the Egyptian Mau. Cats in ancient Egypt were highly respected and even worshipped, and this incredibly beautiful creature (named after its reputed country of origin) is currently the only naturally spotted domestic pedigree feline. Because of its unique qualities, this breed of cat has earned the rather dubious distinction of being one of the most "stolen" breeds when left outside unattended. The Egyptian Mau is an adventurous, outgoing cat, and likes plenty of attention from his owner.

JAPANESE BOBTAIL

A great companion, the Japanese Bobtail, whose roots can be traced back to the country of origin, Japan, is a delightful cat to have around anyone's home. The breed has a small pompom-like tail, but no two tails are ever alike, thus adding to the overall qualities of distinction.

The Japanese Bobtail is in heaven when hanging out with his favorite humans. A friendly, intelligent feline, this cat likes to be kept busy, and adjusts well to practically any environment, including households with children and other pets.

KORAT

The Korat is a rather rare find, and highly coveted by his proud owners, due to a distinctive silver-blue coloring. With an origin in Thailand, the Korat is thought to bring the Thais good luck. This cat is very affectionate and loving, and prefers a quiet evening at home to a night out on the town.

Although it is an energetic breed, the Korat does not like to be alarmed by loud noises and is easily startled. A rather quiet breed by nature, it is happiest in a tranquil environment.

The Maine Coon is known as America's native longhaired cat. It was originally developed as a working cat that had to be able to fend for itself in the woods under extreme weather conditions.

BREED AND BEHAVIOR

The head of the Korat is heart-shaped with smooth, curved lines. The eyes are round when they are fully open, and when they are partially or fully closed they have an Asian slant.

BREED AND BEHAVIOR

MAINE COON CAT

Once upon a time, before the Persian arrived on the show circuit scene, the Maine Coon Cat was the most popular registered feline. After the longhaired Persian hit the US, the revered Maine Coon fell into second place, where it remains today in the world's largest registry, the Cat Fanciers' Association (CFA).

Maine Coon Cats can grow to be one of the largest cat breeds, yet their gentle and easygoing temperaments make them sought-after additions to any family. Despite the myth that this cat is part "raccoon," because of its fluffy, raccoon-like tail, there isn't much of a wild side to this good-natured and "laid-back" feline, who readily adapts to most any situation.

MANX

This is one of the few purebreds to be mated with other breeds of cat to produce the Manx. A Manx, which is known for lack of tail (either "stumpy" or "rumpy"), must be bred with a cat that has a tail, to produce the perfect Manx. However, there are Manx cats that do have a tail, which can be used as breeders, although they cannot compete in the pedigree division of the show circuit.

The Manx is normally a playful cat and exhibits an affectionate disposition. Sometimes, however, a Manx will bond with his owner and have difficulty adjusting to a new owner if there is a change in households. The Manx will never bore you with his many jumping tricks and feats performed with those powerful hind legs.

NORWEGIAN FOREST CAT

The Norwegian Forest Cat could easily be

The head of the Manx is round and slightly longer than broad. Its eyes are large and round and its ears are wide at the base.

confused with the Maine Coon Cat in appearance. Like the Maine Coon, it is a large, muscular cat with a flowing coat of luscious fur. This unique cat, surely built with the help of Mother Nature, has a waterproof coat and plenty of undercoating used to help protect him from the cold, blustery days in his Scandinavian homeland.

This is a feline who loves to hunt, so make sure he has plenty of catnip mice to pounce on in his spare time. And, although the Norwegian Forest Cat has been called an independent kitty, he also likes human companionship and looks forward to daily pettings and special bonding time with his owner.

OCICAT

In the 1960's, when a breeder tried to produce a pointed Abyssinian by mating a Siamese with an Abyssinian, an assortment of kittens was born, one being an unusual golden-spotted male that looked exactly like a baby ocelot. The name Ocicat was derived from that comparison.

This beautifully spotted cat shares many characteristics of the Abyssinian. He is a very social cat that hates to be left alone, and craves human companionship almost constantly. Because he is extremely active, the Ocicat strongly dislikes being confined in small areas, such as apartments, and should be kept in larger areas with room to run at will.

ORIENTAL SHORTHAIR

If you own an Oriental Shorthair feline, more than likely, you are used to hearing him "talk" to you

BREED AND BEHAVIOR

The Ocicat is a hybrid breed that developed through crossing the Abyssinian, American Shorthair, and Siamese. It has been named for the South American wild cat, the ocelot.

Known for his sweet expression, the Persian's personality is built to match his pleasant looks. Persians are quiet and one of the more inactive breeds of cats. Not that it doesn't enjoy spontaneous bouts of play—just not as often as most other breeds.

Persians make up over two-thirds of the registered felines in the Cat Fanciers' Association, so they must own a variety of endearing qualities to make them such a popular choice for cat lovers. If you want a decorative piece of royal grandeur in a thick, fur coat, get a Persian. In addition to their obvious beauty, they are very affectionate as well.

RAGDOLL

As the name suggests, this breed of cat loves to be held, and literally becomes a "ragdoll" in the arms of his owner. A Ragdoll won't

The Persian is the most popular breed in the world. They are believed to have descended from the Turkish Angora.

on a regular basis. This is a particularly vocal cat, to such an extent that, when he wants his own way, he will almost insist upon it until you eventually give in to his demands.

In terms of appearance, it merely looks like a colored or patterned Siamese. His personality is also consistent with that of the Siamese breed, in that the Oriental Shorthair is profusely affectionate and does not require an abundance of "space." The more time he can spend with you, the better—and quieter—the entire household will be.

PERSIAN

The Persian is about as laid-back as a cat can be.

demand much from the owner because it is one of the most quiet and gentle breeds around.

This easygoing feline has a loving disposition and is not the least bit aggressive. Much like the Persian, he is somewhat of a showpiece, and loves to be

In the US, the Oriental Shorthair is a self-colored, shaded, smoke, or tabby Siamese. In the UK each self-color is recognized as a separate breed.

BREED AND BEHAVIOR

The most outstanding characteristic of the Russian Blue is the double coat, which is silky, stands upright, and is short. The coat should have silver-tipped guard hairs to contrast against the solid blue body color.

BREED AND BEHAVIOR

The Ragdoll is a wonderfully docile animal that shows a lot of affection toward its owners. It is a large cat that can take up to three years to mature fully.

gently caressed and cuddled. The Ragdoll is a rather meek cat and prefers quiet and peaceful surroundings.

RUSSIAN BLUE

Once you have owned a Russian Blue, perhaps no other cat will ever equal in comparison. This silver tipped blue cat has a striking presence, and is a most graceful creature. Although a Russian Blue can be playful at times, he also enjoys times of quiet solitude.

This is a feline who usually attaches himself to one human and remains loyal and devoted to that person for the duration of his life. But, unlike some other breeds, he can be left alone without destroying your home or sulking about the place when you return. They are good-natured and loving, and a perfect addition to any family.

SCOTTISH FOLD

All Scottish Fold kittens are born with straight ears, and which kittens will develop folded ears and which will remain "normal," or straight, will not be evident until they are approximately one month old. Then, the "true" Scottish Fold kittens appear before your very eyes.

First found in Scotland, this folded-eared cat can hear perfectly well and will fill your home with happiness and affection. With his big, innocent, round eyes, this sweet-faced cat will demand nothing from you except plenty of love.

SIAMESE

As evidence of its popularity, anyone—even the novice—can distinguish a Siamese cat. The telltale points and elongated body are hard to miss and, to be sure, it is impossible to miss his distinctive voice!

The Scottish Fold first occurred as an ear mutation in some farm cats in Scotland. The breed was established by crossing the British Shorthair and various domestic cats of both Scotland and England.

BREED AND BEHAVIOR

The coloring of the Siamese's short, fine coat is called pointed, meaning that the mask, ears, feet and tail are clearly defined and consistently marked.

The Siamese is the most vocal cat in existence, and his loud, demanding meows can be heard above most other sounds. When your Siamese speaks, you had better be listening!

This relentless feline loves his owner with unequaled passion, but wants his own way no matter what. With a Siamese around, life is definitely not boring.

SOMALI

Basically speaking, the Somali is merely a longhaired Abyssinian. Like his shorthaired counterpart, this feline is quite intelligent and requires mental stimulation to keep him from becoming bored. The Somali always has his own agenda and, if kept active, will make a delightful companion.

SPHYNX

If you are one of those people who loves cats but does not love the shedding that accompanies the species, the Sphynx breed would be the perfect choice. Although many people do not appreciate the hairless body (which makes it quite an odd-looking creature), Sphynx owners quickly fall head over heels in love with this unique animal.

The Sphynx is not a shy cat and likes a lot of attention and affection. He is basically outgoing and energetic but, like most felines, can be mischievous at times as well. His unique look always makes him the object of conversation during parties.

TURKISH VAN

Most likely a descendent of the once-popular Turkish Angora, the Turkish Van has the unusual distinction of being a cat who actually likes water. Normally, cats and water do not mix at all, and trying to get your cat into a tub full of water is nearly impossible. But the Turkish Van will be more than delighted to go skinny-dipping in the pool and swim a few laps like a star Olympian. Besides his predilection for water, the Turkish Van is an agreeable, sociable cat who will provide you with years of entertainment.

Although the Turkish Van is a very independent cat, it is still highly affectionate toward people.

BREED AND BEHAVIOR

35

The Somali is the longhaired version of the Abyssinian. It developed due to a spontaneous mutation, and it possesses all the personality and charm of the Aby.

COMMON KITTY QUIRKS AND HOW TO HANDLE THEM

As a cat owner, you should be aware of your responsibility for your cat's safety and well being.

COMMON KITTY QUIRKS AND HOW TO HANDLE THEM

If you own a cat, or have ever lived with one, you undoubtedly realize that felines have many strange habits and bizarre behaviors that are completely irrational to the logically thinking mind. Nevertheless, the following are just some of the many quirks and curious characteristics that have been associated with the feline population.

BED-SWAPPING

Your cat insists on sleeping in your favorite chair. No matter what cozy alternatives you offer her, she always ends up in that chair. So, you decide that she wins and you get a fluffy blanket or soft pillow to put in the chair and, next week, she won't go near it. Instead, she insists on sleeping on the corner of your sofa. Bed-swapping is a favorite pastime for bored cats. Just when you resign yourself to the fact that your cat will stay in a certain spot, she changes it! There is nothing you can do about this practice except be very careful where you sit!

CARPET SCOOTING

Anyone who has ever witnessed the rather disgusting act of watching your cat drag or "scoot" her rear end across the carpet, will certainly never forget it. There are many reasons cats do this, including swollen anal glands, overweight cats unable to reach their private parts, and just plain laziness.

A trip to your veterinarian should be your

Cats will often find a favorite place to relax and sleep, and then change it! Bed-swapping is a common behavior that many cats have. The only way to deal with this is to be careful of where you sit.

COMMON KITTY QUIRKS AND HOW TO HANDLE THEM

A sit in! Although it looks like this cat may be protesting the painting that's going on, he really just thinks it is a comfortable bed!

first step in order to make sure your cat is not in pain. If your cat is too fat to clean herself, talk to your veterinarian about putting her on a diet. Finally, if you have a feline who is too lazy to hike up that leg and give herself a good washing, you'll probably just have to hope that she won't decide to "carpet scoot" when you have company.

CARPET SHREDDING

It is no secret that carpeting is expensive. That's why a cat who likes to shred carpet is a frustration to his owner. Some cats will dig at a carpet by a closed door, while others just seem to like to dig, especially if they find a spot where something has been spilled in the past.

There are several remedies to this problem. If there is a particular place your cat likes to dig, try cleaning it thoroughly and then place something over that spot for a while. Another deterrent might be to try putting plastic over the area your cat tries to shred. Finally, make sure your cat's favorite scratching post is not made of carpet because that could send a mixed signal in regard to carpet, generally. A good alternative to carpet-covered posts is the sisal (rope) variety.

CATNIP ROLLING

Catnip is meant to be eaten and sniffed—but rolled in? There are indeed

Some houseplants can be poisonous, so use caution in this regard when it comes to your selection of greenery for your home and patio.

cats who love to roll in catnip! There is nothing really wrong with this behavior except that the

Catnip given in small quantities, as pictured here, will not allow your feline friend to roll about in it.

COMMON KITTY QUIRKS AND HOW TO HANDLE THEM

catnip-covered cat will probably become very popular with other felines he encounters.

CHRISTMAS TREE CLIMBING

Have you ever come home to find your carefully decorated Christmas tree tipped over, and many of your favorite ornaments

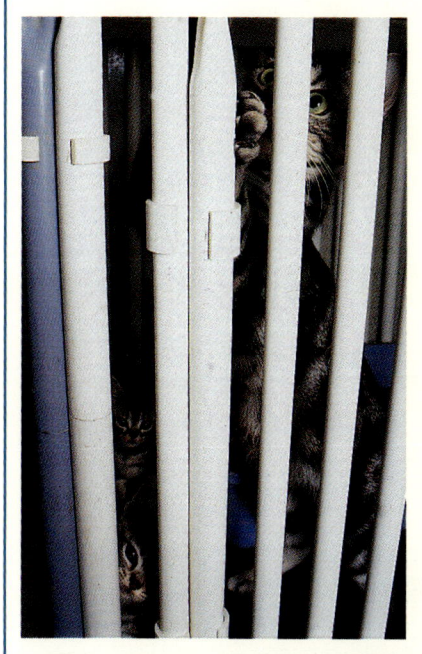

Many cats do not like being locked inside a crate. They usually try to dig, scratch and howl endlessly until they are let out. This is very much the same behavior one will display when locked out of a room. Nothing drives a cat more crazy than having a door closed to a room that it wants to enter.

broken? If you have a cat, this is a very real possibility. Especially if you have a real tree instead of an artificial one. To a cat, trees hold one purpose, and this is, of course, to be climbed! They do not understand our odd tradition of decorating a perfectly good climbing post simply for the purpose of ornamentation.

First of all, anchoring and tying your tree is a must. A hook in the ceiling and some transparent fishing line will do a fine job and not create an eyesore. This will keep the tree from toppling over. Also, discourage your cat from going near the Christmas tree. If all else fails, do not allow your cat in the same room with your Christmas tree.

Cats know that trees grow so that they can climb in them, and why shouldn't they?

CLOSED DOOR SYNDROME

Nothing will drive a cat crazier than a closed door! As we all know, cats are the most curious of creatures, and not being able to see what is on the other side of a door is pure torture for them. They will

Areas of potential danger, such as the top of a stove, should be off limits to your cat.

COMMON KITTY QUIRKS AND HOW TO HANDLE THEM

If you don't provide your cat with an appropriate area in which it can scratch its claws and play, it will select a spot on its own that may be less to your liking.

dig, scratch, and howl endlessly until the door is opened. Then, once they are let in to the "secret" room and the door once again closes, the process starts all over again. They seem to have mysteriously forgotten what is on the other side of the door they just entered.

It would be nice to say there is a solution to this all-too-common problem. You can ignore the cat, and hope she gets tired of howling, scratching and digging. (Good luck.)

Everyone thinks cats tire easily because they sleep 16 hours a day, but when it comes to a closed door, they are on constant alert. Until science comes up with a better plan, the most expeditious solution to this problem is to just open the door!

CORD CHEWING

Dogs are known for their chewing habits, but occasionally, a cat will also develop the dangerous cord chewing habit as well. Chewing on electrical cords and wires must not be permitted because if your feline is able to chew through it, she will surely be electrocuted. So, you must do everything in your power to prevent this practice. Some people will put foul-tasting substances, such as Tabasco sauce or cayenne pepper, on electrical cords, so that the cat will take one taste and decide cord chewing is not for her. You can also tape wires so they won't be exposed or, as a

Your cat will want to investigate every area that it can get to. Your library or favorite bookcase will not be excluded from his inquisitiveness.

COMMON KITTY QUIRKS AND HOW TO HANDLE THEM

Pet doors are specifically designed to allow your cat access to the outdoors when he wants to go. This sometimes settles a portion of his curiosity and will make a cat less restless when indoors.

last resort, have your wiring installed within your walls, although this is an extreme and costly measure.

COUNTERTOP AND TABLETOP VISITS

If you often leave food exposed on your counter and table tops, your hungry and curious cat will probably become a regular visitor. No one likes to see an animal on their dining or food preparation areas. The first rule in avoiding this situation is to refrain from leaving any food out while the cat is unsupervised—ever! Also, do not feed your cat scraps or morsels in the kitchen while you are preparing dinner or eating at your table so that your cat will not associate food with the kitchen and dining areas.

If your persistent feline still likes to visit your table and countertops, you might want to put something on these surfaces that will startle the cat when she jumps on them. Empty aluminum cans clanking to the floor, or spoons strategically placed, might just scare her enough to avoid table and countertop visits in the future. Also, cats hate the feel of plastic or tape on their paws, so covering up the areas with these items might deter your curious cat from food and dining zones once and for all.

Scratching posts can be covered with a number of different materials. Some cats prefer a carpeted surface while others may prefer sisal.

CURTAIN CLIMBING

Do your curtains have little cat claw holes all up and down them? If you have cats and curtains,

Counter and tabletop visits by your cat might be cute to you, but your guests will probably not enjoy having your kitty pouncing about these while they visit. Try to deter your cats from these areas by placing things on the areas that may startle him when he jumps there.

COMMON KITTY QUIRKS AND HOW TO HANDLE THEM

Fine china and other breakables are safest in a closed cabinet where your busy kitty is unlikely to knock them over.

you might find that the two do not exactly mix. Cats are natural-born climbers, and they like to be on higher ground (like the top of your curtain rods) so they can watch for approaching enemies.

To stop this behavior, many people simply hang their curtains up by loose thread so that each time their cat climbs them, they will fall down on top of the cat. They will soon learn that these are not "climbable" objects, and a more permanent sewing job can be completed.

Another alternative to curtain climbing is to invest in vinyl mini-blinds or vertical blinds. These window coverings are relatively inexpensive, difficult to climb and, best of all, you won't find cat hair on them.

ESCAPE ATTEMPTS

You open your door to come in the house and see a furry flash zoom by. Your cat has escaped into the great outdoors once again. Then hours of searching and calling ensue until the cat becomes bored or hungry enough to come home on her own—assuming she makes it home safely.

The most tried and true way to prevent this is to have someone hide outside the door and offer one good squirt of water when the cat tries to escape. A loud noise may also deter your escape artist, and startle her back inside. If your cat associates spraying water and loud, unpleasant noises with going outside, she will soon be content to stay indoors where it is safe.

Also, be aware of your cat's presence and don't try to come in the door with an armload of groceries. If you plan to have guests in and out of the front door, put your cat in another part of the house until the guests have left.

FINICKY APPETITES

Morris, the finicky cat, has been a popular television commercial star for years. The belief in the

The great outdoors may entice your housecat to explore. You may wish to attach a bell or other noise device to your cat's collar so that you can locate him as well as warn birds of his presence.

COMMON KITTY QUIRKS AND HOW TO HANDLE THEM

finicky cat is not a myth, however. Your cat may eat the same food every day for years and then suddenly refuse to eat it. Or, the new, improved cat food that is much healthier for your cat has her turning her nose up in disgust. What can you do if you own a feline with a finicky palate?

First, consider a feline's unique nutritional needs. Cats are true carnivores because their bodies need the protein and vitamins that are found in meat. Cats need a lot of protein and very few carbohydrates. They also need certain vitamins and minerals, most of which

Cats with finicky appetites can almost always be enticed to eat if offered a helping of jarred baby food.

come from meat and bones. One thing cats cannot live without is the essential amino acid taurine, something their bodies cannot synthesize on their own.

Today's domesticated cats are fortunate in that cat food manufacturers have developed foods specifically designed to meet the cat's nutritional needs—foods high in protein with the necessary vitamins and minerals as well as taurine. These foods, which come in dry, canned or semi-moist form, include feeding directions to take the guesswork out of how much to feed, though owners are wise to monitor their cats' intakes and consult with their veterinarians should the

cat appear too thin, too heavy or shows other signs of distress.

Equally important to nutritionally complete food is access to fresh, clean water at all times. All animals need this, and cats are no exception!

Another mistake cat owners sometimes make is to give their pets milk. Cats do like the taste of milk, but their systems really can't handle it, and usually cats who drink milk or eat milk-based products will develop diarrhea.

If any of a cat's nutritional needs aren't being met, or sometimes even if they are, the cat may become finicky. It's annoying when you realize the bag of food you just bought no longer appeals

Treats can be provided on an occasional basis to help provide a little variety in the diet. Some treats act as a cleansing agent to help reduce tartar on the cat's teeth. Photo courtesy of Heinz.

COMMON KITTY QUIRKS AND HOW TO HANDLE THEM

Sometimes finicky eaters can be encouraged to eat by observing another cat eating.

to Kitty, but take it in stride and try something else. You could even try mixing some of the old food in with the new so it isn't completely wasted.

Scratching posts and playgrounds are available at most pet-supply stores. Many are multi-leveled and incorporate a number of different activities for your feline to enjoy and get his exercise.

FURNITURE SCRATCHING

A new couch; your antique coffee table; the beautiful hutch your husband built with his own two hands. All of these items are precious. And, all of these items are potential victims of "cat scratch fever!"

Cats need to scratch. It isn't something they do to irritate their owners; it is a behavior they initiate for many reasons: First of all, cats need to shed the sheaths of their claws so they can grow new ones, much in the same way humans continually grow new fingernails and toenails. Secondly, cats will scratch in order to "mark" their territory with scent glands in their paw pads.

The good news is that furniture scratch is one of the easiest "bad" habits to break. One or more sturdy scratching posts should be distributed throughout your home, and, if possible, provide more than one kind of post for your cat—the cardboard type, sisal (rope) type and carpet type offer your cat variety if your cat isn't already a carpet ripper.

More and more experts agree that declawing is a cruel and inhumane answer to problem scratching, and declawing your cat is no longer considered a compassionate choice. Placing catnip on the scratching post will attract your cat, and playing with him near the scratching post helps too. If provided with adequate alternatives, your cat's scratching behavior will soon become a distant memory.

If not provided with a good scratching post, you can be sure your cat will find something that he finds suitable. Lucky for this cat owner it's a tree outside. Many cats seem to take a liking to their owner's furniture or carpeting.

COMMON KITTY QUIRKS AND HOW TO HANDLE THEM

All felines require some green in their diets. Strict indoor cats often dine on their owner's houseplants, which may be toxic. Owners must be wary of this bad habit.

GARBAGE LOOTING

Arriving home after a long, hard day only to find your trash can tipped over and garbage strewn all over the kitchen can make even the most understanding pet owner unhappy. The cat's curiosity overtakes all reasonable doubt about whether he should be digging in the trash—especially if there are tasty leftovers inside. The best way to stop garbage looting and trash tipping is to put the trash can out of sight and out of reach.

Many people keep their trash cans hidden underneath their kitchen sink, which should prevent any pet from garbage looting. If this is not an alternative, you can also purchase trash cans with sturdy lids and add extra weights to the bottom to hinder tipping.

GROOMING ANXIETIES

Some cats will run and hide at the sight of a brush or comb. The reason for this is generally unknown. Perhaps your cat does not like to be forced to sit still while being groomed, or maybe you are brushing or combing too vigorously. Whatever the reason, you can teach your cat to accept—if not enjoy—being groomed.

It is important to put Kitty at ease when grooming her. If possible, start grooming your cat when she is still a kitten to get her used to the ritual. Don't handle her roughly, or yell at her if she squirms. Reassuring her in a soft speaking voice will calm your feline through the grooming process. If you don't make grooming a big deal, neither should your cat.

HOUSEPLANT EATING

It is still uncertain why cats eat grass. Many experts believe it has something to do with disturbing their digestive

COMMON KITTY QUIRKS AND HOW TO HANDLE THEM

It is as important for our feline friends to have healthy teeth and gums as it is for us. Fortunately, maintaining oral care is getting easier and easier for the pet owner. Now there's a taste-free, easy-to-use gel that will keep your cat's teeth clean, reduce tartar build up and eliminate breath odor. Photo courtesy of Breath Friend™/American Media Group.

system so they will expel any hairballs caught inside. Others think a cat must be ill if she eats grass. Whatever the reason, cats confined indoors should have access to greenery or they may start the potentially deadly habit of houseplant eating. Before putting any plants in your house, first find out if the plant is toxic to animals.

Regardless of whether the plant is safe for your cat, you won't want her eating your home decorations. If you do have non-toxic plants in your home, hang them on hooks from the ceiling out of your cat's reach. You can also plant grass seeds and grow greens for your cat to nibble on and this will likely divert her attention from your own decorative plants.

INAPPROPRIATE DEFECATION

A cat who empties his bowels anyplace other than the litter box is usually quite a problem for the owner. It is a feline's instinct to bury its waste in soil or litter, so you don't have to "teach" a cat to use the litter box. Naturally, you should always show your cat where the litter box is located and try not to move it or place it in a heavy traffic area of the house.

It's important to keep your cat's litter box clean, and a variety of fast-acting, clean smelling litters is available to make your job easier. Photo courtesy of Swheat Scoop/ Pet Care Systems, Inc.

COMMON KITTY QUIRKS AND HOW TO HANDLE THEM

Never rub your cat's nose in his accidents. Cats do not understand this type of punishment. If your cat has stopped using his litter box to empty his bowels, there most certainly must be a reason, and it is the responsibility of the cat owner to investigate.

First of all, have your cat checked out by a veterinarian to make sure he isn't ill. Any changes in litter brands, or even the location of the litter box, might disturb your cat, and cause him not to use the box anymore. The solution is to retrain your feline to use the box again. Put your cat's food bowl in the spot in which he is inappropriately defecating (cats will not defecate where they eat), and if you must, place her in a small room with nothing but food, water, and her litter box so she will learn to use the box again.

Finally, cats are fastidiously clean animals. A dirty, smelly litter box will most definitely make Kitty look for cleaner facilities. Anything and everything you can do to encourage your cat to use his litter box should be considered a priority in order to keep everyone in your home happy and content.

INAPPROPRIATE URINATION

Nothing has a smell quite like it. Cat urine has an odor much like that of harsh ammonia. Cat litter

Above: Cats are fastidiously clean creatures who can be fussy about their litter boxes. Cat owners need a litter that absorbs odor and is easy to manage. One made from wood and bark by-products can meet the satisfaction of cats and owners alike. Photo courtesy of Gentle Touch Products. **Below:** Litter box overhang is often an accidental event that most cats are not even aware they do. A bigger litter box may be the solution you are looking for to prevent this from happening.

COMMON KITTY QUIRKS AND HOW TO HANDLE THEM

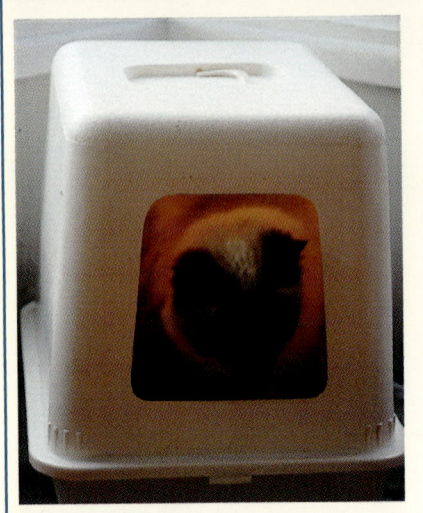

A covered litter box may prevent litter box overhang from happening.

is designed to soak up the smells of cat urine and absorb it. Suppose your cat isn't using the designated bathroom area. And, suppose as well that your cat urinates somewhere other than the litter box?

The main reason cats stop urinating in their litter boxes is because they are ill, usually with a urinary tract infection. By urinating in inappropriate locations, your cat is trying to get your attention. If it hurts to urinate, she obviously cannot tell you verbally. Therefore the only way she can get your attention is by changing her habits and making you aware that something is indeed very wrong. Do not punish her for her actions. If your veterinarian gives her a clean bill of health, you will then have to look for behavioral problems.

Any changes at home, including new household members (animal or human), a change in residence, or even a dramatic change in your schedule, can make your feline develop behavioral problems. Observe your cat, and make every effort to discover the source of her troublesome conduct so that together you can solve the problem.

LITTER BOX OVERHANG

Litter box overhang is actually an accidental event that cats are likely unaware of. It occurs when a cat climbs into his litter box to do his business and, even though all four paws are stepping in litter, his rear end actually hangs over the side of the box, outside of the litter area. Thus, the cat unknowingly goes to the bathroom on the floor but, because his feet are in the litter, he thinks his whole body is in the litter box.

How do you prevent this? One solution is to get a bigger litter box because the current box may be too small for the size of your feline. And, make sure you keep the box clean because Kitty might be standing near the edge of the box because the litter is soiled elsewhere in the box.

The easiest way to prevent this habit is to purchase a covered litter box so the only place they could overhang would be the entrance opening (leaving less chance of overhang because a cat normally won't go into an enclosed structure and leave the back open to possible predator attack). Then, the cat will go in the box and turn around so that he is looking out of the hole. Problem solved!

LITTER FLINGING

You sometimes have to wonder if the art of "litter flinging" is not achieved through years of practice. This occurs when your cat covers his deposits in the litter box, and digs with such gusto that the litter flies out of the box and all over your clean floor.

Investing in a covered litter box is one way to contain this habit, although, with a little practice, your cat's aim may improve. Providing

Kitty litter has come a long way since the days of dry clay! Today's litters are dust free, super-absorbent and don't get stuck in kitty's paws and then end up in your carpet! Best of all, cats are comfortable using them. Photo courtesy of Yesterday's News®.

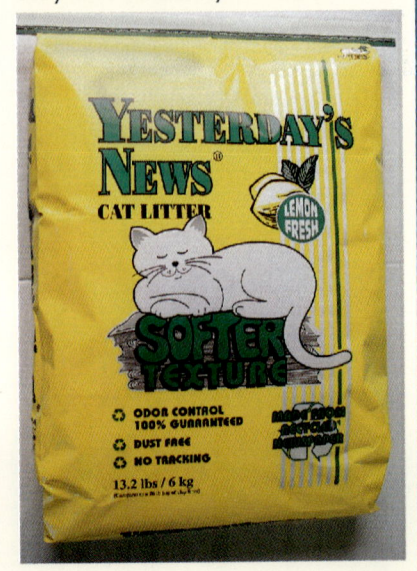

COMMON KITTY QUIRKS AND HOW TO HANDLE THEM

Your cat's playing area and all its surroundings are considered his territory. Sometimes cats become very possessive of their areas and, if unneutered, a tom cat will spray, or mark, his territory. The only way to curb this type of behavior is to neuter your cat.

some sort of tray or mat underneath the box to catch all the "flung" litter is another alternative.

SPRAYING/MARKING TERRITORY

Spraying is the act of a cat marking his territory. Although any cat, male or female, altered or whole, can take up this smelly ritual, most cats who do spray are unaltered or whole male cats. Male cats are driven by their hormones, and thus marking is an instinctive behavior used to establish territory. Sometimes, however, the instinct becomes a habit.

The most obvious solution to your cat's spraying problem, if the feline is a tom cat, is to have him neutered. Unfortunately, if he is old enough to have begun spraying, he may never stop. You must also have your veterinarian check your cat for any medical problems that may be causing spraying, such as FUS or cystitis. It is also important to keep the litter box very clean and make sure there is a separate box for each cat in your home. Removing the odor with a good (non-ammonia) cleaning solution is also advised to help prevent similar incidents.

TOE ATTACKING

It is 3:00 a.m. and you are in your warm, cozy bed having a really great dream about winning the lottery. Suddenly, your rewarding moment is painfully

If your kitty enjoys attacking your toes or plays frantically with the toilet paper throughout your home, give him other things to chase and attack. Many fine cat toys are available to satisfy these habits from your local pet store.

No matter how tidy and trustworthy your cat, accidents will happen. All you can do is be prepared by being armed with a non-toxic, effective stain and odor remover. Photo courtesy of Francodex.

interrupted by the sensation of gnashing claws and teeth sinking into your big toe. Your response is to jump straight up in the air, and reach for the lamp, only to find your faithful cat at the end of the bed in search of her flesh-colored toy.

Your only alternatives are to lock her out of the bedroom (which can cause other problems), sleep without ever wiggling your toes again or, my personal favorite: wait until your cat falls asleep to win your rightful revenge by grabbing her back right toe and interrupting *her* glorious dreams!

TOILET PAPER CAPERS

Have you ever gone into your bathroom and found the roll of toilet paper shredded into confetti? If so, your cat has performed the oldest trick in the feline

COMMON KITTY QUIRKS AND HOW TO HANDLE THEM

If your feline is an undercover kitty, be careful before you climb into bed. Check under your covers before you lie down so that you do not hurt or injure your pet.

Your feline friend will be with you for a great many years, and it is therefore up to you to provide him with everything he needs to keep him content.

book—the ever-popular toilet paper caper! Your cat has just discovered that rolls of toilet paper go around and around and it feels really good to sink his claws into it.

There are several remedies to this problem, including not allowing your cat into your bathroom, or placing the roll so that the toilet paper comes underneath instead of over the top—although a smart cat will probably figure this out in no time! Or, you may have to invest in one of those plastic toilet paper covers intended to stop the famous paper caper once and for all!

UNDER "COVER" KITTY

Many cats love to be covered up. Whether it is a throw rug in the den or the bedspread on your bed, your feline may decide that a warm, dark place under the covers is perfect for a mid-afternoon nap.

There is nothing unusual—or wrong—with this behavior. The only possible danger is jumping into bed or running over that throw rug only to find that the lump you've just encountered is your cat! So, if your cat is an "under-cover" kitty, be careful where you sit, step, or lie down because that rolled-up lump might just be alive!

Cats continually seek warm spots in the house for their naps. Offer a nice soft bed in front of a sunny window or else you may find your cat sitting in an open clothes dryer!

COMMON KITTY QUIRKS AND HOW TO HANDLE THEM

Many cats decide that a warm, dark place under the covers is a perfect place for a mid-afternoon nap.

WARMTH SEEKERS

Since cats have a normal body temperature of approximately 102 degrees (significantly higher than the human norm), it is no wonder that they constantly seek warm spots in the house for a cat nap.

Beware of your cat's discovery of an open clothes dryer because she might just cuddle up inside without you knowing it, and become seriously injured (or even killed) if you decide to dry some clothes. It is fine for your cat to seek out warm spots, as long as you make sure they are not "dangerously" warm ones.

WINDOW SILL SITTING

For a cat, window sill sitting is the next best thing to being there. Your cat needs access to a window (or windows) with a good view. Nothing passes the day faster than watching birds, squirrels, and chipmunks scurrying busily out in the yard. And, don't feel sorry for your cat because she is not outside catching these little creatures. It is much safer for your cat to be kept away from natural outdoor predators. Simply provide your cat with a comfortable seat—preferably in a sunny spot—to spend the afternoon gazing out at the world. You can make or purchase sturdy window shelf seats designed especially for cats.

If any of the behaviors mentioned remind you of your feline companion, remember this basic tenet of cat behavior: It is far easier to work with your cat than against him. If you take the time to figure out why your feline is exhibiting these sometimes aberrant, yet always fascinating, behaviors, you will be well on your way to understanding the most intriguing and mysterious creature known to mankind—your charming, cherished cat.

Fish tanks often prove frustrating to cats, as they truly dislike water, but really want what's inside. They are intelligent, inquisitive animals, and nothing escapes their attention.

CATS AT PLAY

No matter what the age, breed, gender or background of your favorite feline, one thing is for certain—cats will always find time to play! With those mischievous peek-a-boo looks and those swaying, swishing tails, it seems that kitty can always find a spare minute, day or night (but usually night), to take a wild run or merry romp through the house.

Whether your frisky feline is chasing her shadow, one of her littermates, the family dog, or merely her own tail, playing is an essential part of the cat's overall personality. But why is it essential for you to ensure kitty makes ample time for "playtime" each day? For starters, playing provides your cat with much-needed exercise as well as an inexpensive form of entertainment. Remember, too, that a bored cat is usually a naughty cat!

UN-FRISKY SOMETIMES EQUALS UN-WELL

You won't usually have trouble finding a cat that doesn't want to play at all. Of course, some older cats or sick cats may appear to have no interest in playing. Sometimes the lack of play can even alert the unsuspecting cat owner that kitty is not well. Many times, the only change in an ill feline will be the absence of its usual "playtime."

Owners may take their

This little kitten doesn't know what he should play with first. Kittens should not lack playfulness or friskiness.

CATS AT PLAY

Kittens learn the shape and consistency of objects by pawing at them.

"My cat doesn't ever want to play anymore." An examination and perhaps a few tests will usually find that something such as arthritis (in older cats) has set in or maybe even a disease or virus has taken away kitty's desire to frolic. A lot of cat owners have discovered their cat's illness simply from observing kitty's activity level (or lack thereof).

A playful cat, then, is usually a cat that is feeling good about herself and wanting to sharpen those stalking and pouncing skills taught early in life

The lack of activity in your cat could be the first sign that he is not feeling well.

furry friends to the veterinarian with basically no symptoms except for

(during the critical early learning and socialization period) by virtually every

This looked more comfortable than it actually is. When it comes to your kitten's undying curiosity, nothing is sacred.

CATS AT PLAY

Your cat can easily keep himself occupied, but he will greatly enjoy spending quality playtime with you.

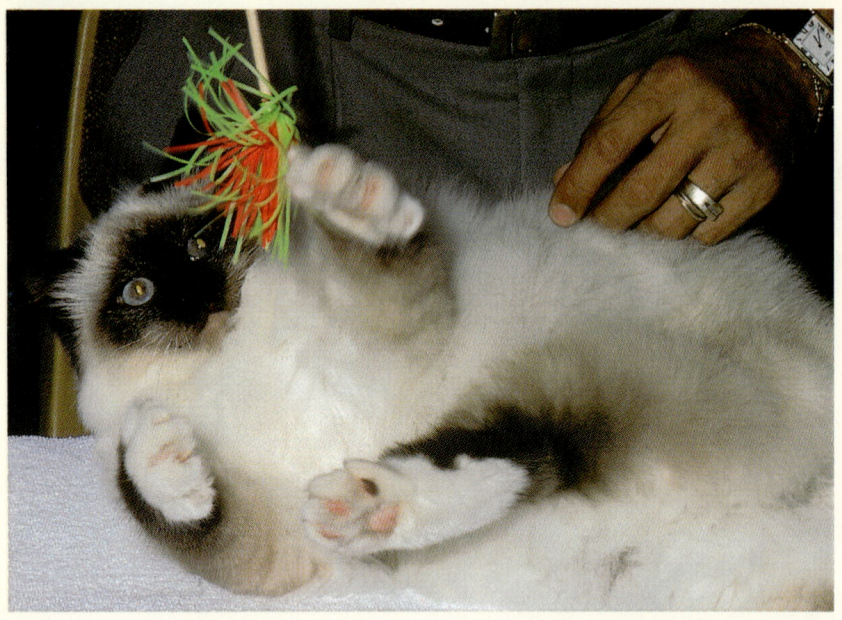

Your cat will enjoy playing with a variety of toys, especially those that are small enough to be batted around.

mother cat. It is so important that these instincts and desires not be ignored by cat owners. If you have more than one cat (or maybe even a dog that kitty gets along with very well and would never harm the cat...even while innocently playing), your job may be somewhat easier because a lot of those needs will most likely be met by your cat's companion.

ROUGH HOUSING TOO ROUGHLY

You should be concerned, however, if "playtime" ever takes a dangerous turn or seems to be getting too rough. Then, you will need to intervene—especially if one of the cats (or whatever type of pet is playing with your cat) appears to be the "bully" or frequently ends up hurting more than just kitty's pride (which is a very delicate thing, you know)!

Also, when you have been rough-housing and playing with your cat, you should be able to recognize the warning signs that the cat's playing, at some point, has become an act of aggression or true anger. Hisses and growls are not part of the pattern of usual cat play behavior. When your feline's ears go back against the head and kitty starts growling and/or hissing, this is a sign that whatever behavior you are exhibiting toward your cat (whether it be tickling or some other form of teasing) you had better stop immediately if you don't want to get bitten or scratched.

No, your cat does not *want* to hurt you but you must remember that the feline has not been domesticated as long as the canine (and even the friendliest of dogs can turn on its owner as well) and those wild instincts still rage within even the tiniest of domesticated kitties. No matter what size your petite feline is, he probably

Two buddies engaged in a harmless tussle. If you own two cats, they can keep each other company when you are not home.

CATS AT PLAY

When no one is around to give kitty a back rub, the cement flooring does nicely.

sees himself as a seven hundred pound tiger when he looks in the mirror (especially when his fur is fluffed to the max and standing on end during certain intimidating circumstances)! So, remember, when the playtime turns rough with another animal, an abrupt loud noise will usually stop the fighting and, if you suddenly become the target of a feline's attack mode, the smartest move is to slowly retreat—in other words, tuck your tail between your legs and head for the hills!

If you choose to leave your cats outdoors while you are not home, be certain they have some sort of shelter they can retreat to in case the weather gets bad.

KITTY PARAPHERNALIA

Kitty paraphernalia is classified as those "extras" you should have around to make absolutely sure that your overly resourceful feline has his own "things" that will help to occupy his complex mind.

For starters, it's not just a good idea to have a scratching post to save such prized possessions as your antique coffee table and brand new plush L-shaped sofa, it is practically a necessity. Cats love to scratch, and since the painful practice of declawing felines is becoming more and more uncommon (thank goodness), cat owners want to find a way to prevent their cuddly companions from shredding every piece of furniture in the house.

The solution is a simple one—a scratching post. Not only is it an excellent form of exercise for kitty (stretching muscles, etc.) it will also help the cat to get rid of the dead parts of the nail, called the sheath, effortlessly. It is a common misconception that felines scratch their owner's furniture to get attention or to get their owner "mad" at them or to seek out "revenge" because of a recent public scolding

The exercise your cat will receive from playing outdoors is great, however, the number of dangerous situations he can get himself into outweigh any possible benefits.

CATS AT PLAY

Houseplants are very enticing to cats. They love to play in the dirt as well as eat the foliage. Many houseplants are poisonous, so use caution in this regard when it comes to your cat.

CATS AT PLAY

during a family dinner party.

Felines do not, however, understand the art of revenge (although it would appear they do on many occasions). Cats scratch because they NEED to. It assists in the shedding of the sheaths, exercising of their legs and feet and, also, a set of scent glands are found in a feline's paws so it is also a territorial marking procedure. That is why you should invest in a sturdy scratching post covered with carpet, sisal rope, or even just a plain wooden one. Your cat will most likely show a preference to one type of post, however, and that is

Encourage your cat to play near his scratching post so that he will spend his clawing time there.

the type you should then have on hand regularly.

You should encourage your cat to play at the scratching post. If you need to dangle a toy around the post or manually put kitty's paws on the post and show the scratching motion, then do so. Once your cat learns that the scratching post is his and his alone to use as he sees fit, your furniture, walls, curtains, and trimboards will all have a longer survival rate. Plus, he will be a much happier kitty because he's found a place of his own to play.

CATNIPPING AROUND

Okay, you have the ONE cat that refuses to go near that unattractive, artificial-

There's no better place to rest than in front of a bright, sunny window. These two pals are busy soaking in the sun.

CATS AT PLAY

Catnip can turn even the grumpiest of cats back into playful kittens once they get a taste of it. In fact, many felines even seem to get downright giggly and giddy after indulging in a little of it.

looking, fake tree (also known as the scratching post). Dangling strings won't do the trick; strategically positioning kitty near the scratching post only ticks her off more. You need a secret weapon—CATNIP!

A lot of felines (but not all) seem to get downright giggly and giddy on the cat world's answer to a swig of grandpa's secret recipe for moonshine—the green-growing grassy plant known as catnip. An object that your cat once ignored (like the scratching post or even a new toy you bought which she's snubbed since you unwrapped it for her) might just become irresistibly tempting with a little dose of catnip added to it.

Even the grumpiest of cats have been known to turn almost "kittenish" once again after a taste or smell of the good stuff. Many felines can get "high" on it by smell alone while others like to eat the dried plant (catnip is sold as dried in stores but it is possible to raise your own catnip plants and give the freshly grown treat to kitty). Whichever way your cat decides to use catnip, if it does indeed have an effect on kitty, you will

"I know there's some catnip around here some place." Many felines go crazy for catnip and will come from their hiding place at the sound of its package opening up.

Catnip stimulates the central nervous system in cats. Cats differ in their responses to catnip: head shaking, drooling, head rubbing, rolling glazed eye and hallucinatory behavior.

CATS AT PLAY

Daytime already! Cats are creatures of the night and love to play and hunt during the late-night hours.

soon know that it worked!

Although it affects all cats differently, most have been known to stalk and attack anything that moves or, for that matter, doesn't move. Running without going anywhere as well as jumping to incredible heights are also sign that a "nip" of the stuff has had the desired effect. The good news, however, is that a hyperactive, playful cat soon becomes a worn-out, *very* sleepy cat.

CREATURES OF THE NIGHT

Unless you work the graveyard shift, you probably use the nighttime hours of the day to get your eight hours of sleep (or anywhere from three to sixteen hours, depending on the individual). This is the time of day that your body unwinds and prepares for peaceful

Just when you are ready to go to bed for the night, your cat will be ready to romp and play. This British Shorthair is just waiting for something to come along that he can pounce on.

Other than the nocturnal, cats have *nothing* in common with opossums and skunks!

slumber. Upon the rising of the sun, a new day begins and we, as humans, wake up to face it.

But if you are of feline origin, the sun going down means something different—it's time to party! Known as the evening or eleven o'clock crazies (the time, eleven o'clock, seems to be the time most humans get ready to go to sleep), this is the time that cats decide is the perfect time to get rowdy. In other words, when we start winding down, our frisky felines begin to get wound up—literally!

Usually it begins by seeing a furry streak run by you. You can't be certain, but you think it was your cat going by at top speed. "Great!" you solemnly think, "Just when I'm ready for bed, kitty wants to play." It never

fails, as almost any cat owner will tell you.

The reason for the evening crazies is because cats are nocturnal creatures or, mind you, creatures of the night. Just as many wild animals such as opossums, skunks, raccoons, and bats use the nighttime hours to explore, the feline uses the nighttime hours to play. Have you ever had a cat jump on your head at 4:00 a.m.? Or how about two cats (or a cat and a dog) trample across your bed like a herd of elephants just when you're dozing off into a fantastic dream about winning the lottery? It is, if nothing else, a very rude awakening.

Of course, you *could* make your bedroom off-limits to the pets of the household but, as any cat owner will attest to, a

CATS AT PLAY

closed door only makes matters worse. Meowing, scratching, digging, and other loud, disturbing displays of unsatisfaction at the sight of a closed door will keep you up much longer than just keeping the door open and using your clever human wit to ensure you will not be bothered by a playful kitty while you are trying to get a good night's sleep.

A cat, for some reason or other, has a very short attention span. Unless he is waiting for that mouse he saw come out from behind the refrigerator (felines have been known to wait on a mouse to reappear from such places for hours, possibly even days), then it won't take too much exertion to convince kitty it is time for a very crucial act—a much-needed catnap.

So, about 20-30 minutes before you prepare to take your long winter's nap, it would be an excellent idea to make this the nightly "playtime" for both you and your kitty. After about 10 or 15 minutes of you tossing stuffed mice by your cat's head and dangling a string just out of her reach so that she has to jump and jump and jump to catch it, somebody other than YOU is going to be very tired.

Congratulations, you've just accomplished an important goal. You've gotten your cat to exercise, to use her stalking and hunting instincts, and you've challenged her to the point of complete and utter physical and mental exhaustion. Now everyone in your household, humans and animals alike, can take a well-deserved catnap.

Taking time to play with your cat will benefit everyone involved and ensure a home that's a happy and healthy one for all who dwell within.

Expend some of your cat's energy by playing with him for a short while before you go to bed. Your cat will exhaust himself and will also look forward to a nice long rest.

SUGGESTED READING

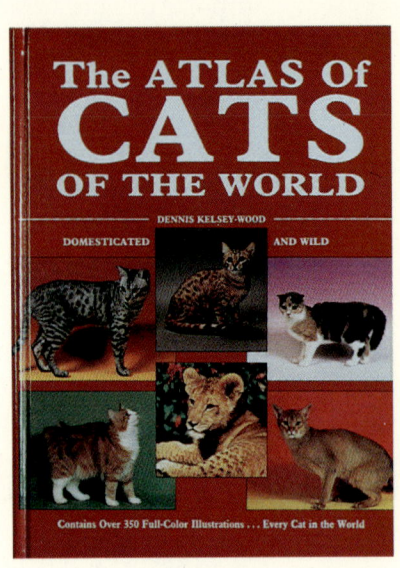

TS-127
384 pages,
over 350 full-color photos.

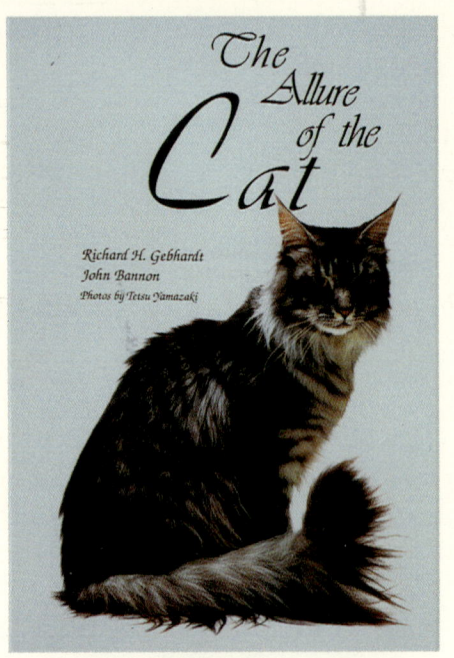

TS-173
304 pages, over 400 full-color photos.

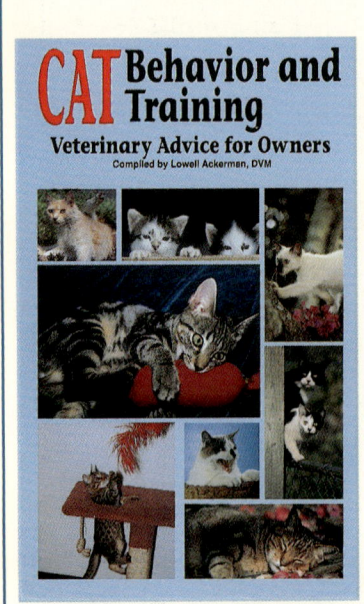

TS-253, 320 pages, over 230 full-color photos.

TS-219, 192 pages, over 90 full-color photos.

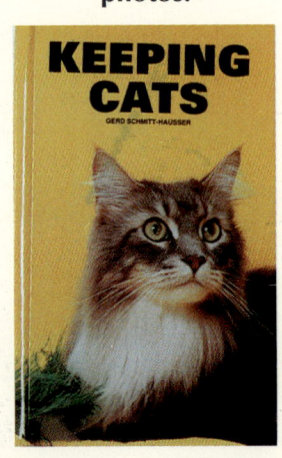

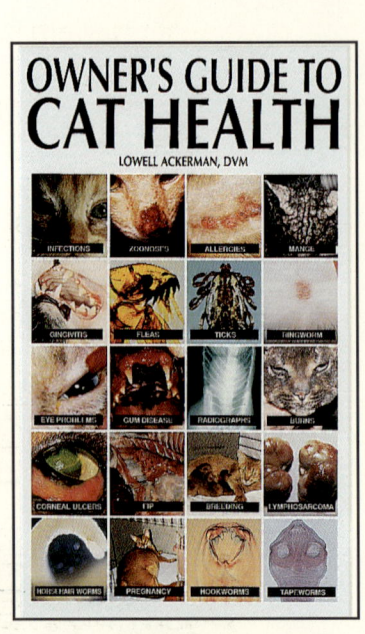

TS-251
208 pages, over 160 full-color photos.

TW-103
256 pages, over 200 full color photos.

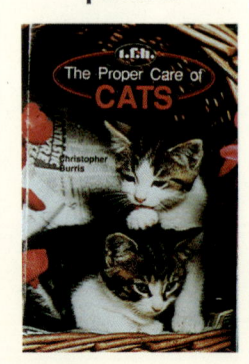

Acknowledgement

This volume in the Basic Training, Caring & Understanding Library series was researched in part at the Ontario Veterinary college at the University of Guelph in Guelph, Ontario, and was published under the auspices of Dr. Herbert R. Axelrod.

A world-renown scientist, explorer, author, university professor, lecturer, and publisher, Dr. Axelrod is the best-known tropical fish expert in the world and the founder and chairman of T.F.H. Publications, Inc., the largest and most respected publisher of pet literature in the world. He has written 16 definitive texts on Ichthyology (including the best-selling Handbook of Tropical Aquarium Fishes), published more than 30 books on individual species of fish for the hobbyist, written hundreds of articles, and discovered hundreds of previously unknown species, six of which have been named after him.

Dr. Axelrod holds a Ph.D and was awarded an Honorary Doctor of Science degree by the University of Guelph, where he is now an adjunct professor in the Department of Zoology. He has served on the American Pet Products Manufacturers Association Board of Governors and is a member of the American Society of Herpetologists and Ichthyologists, the Biometric Society, the New York Zoological Society, the New York Academy of Sciences, the American Fisheries Society, the National Research Council, the National Academy of Sciences, and numerous aquarium societies around the world. In 1977, Dr. Axelrod was awarded the Smithson Silver Medal for his ichthyological and charitable endeavors by the Smithsonian Institution. A decade later, he was elected an endowment member of the American Museum of Natural History and was named a life member of the James Smithson Society by the Smithsonian Associates' national board. He has donated in excess of $50 million in recent years to the American Museum of National History, the University of Guelph, and other institutions.

Index

Abyssinian, 23, 30, 34
Adulthood, 7, 8
Age, 3, 5
Aging, 10
Alteration, 18, 19, 22
American Curl, 23
American Shorthair, 24
Arthritis, 54
Bed swapping, 37
Bengal, 24
Birman, 24
Body temperature, 50
Bombay, 26
Breed, 3, 23
British Shorthair, 26
Burma, 24
Burmese, 26
Carpet scooting, 37
Carpet shredding, 38
Cat Fanciers' Association, 28, 31
Catnip, 38, 45, 60
Cervantes, 3
Chartreux, 26
Closed door syndrome, 39
Cord chewing, 40
Cornish Rex, 27
Curtain Climbing, 40
Declawing, 44
Defecation, 46
Devon Rex, 27
Diet, 38
Disposition, 7
Early handling, 12
Early learning, 12, 15
Egypt, 27
Egyptian Mau, 27
Energy level, 10
Escape attempts, 42
Finicky Appetites, 42
Furniture Scratching, 44
Gender, 3, 18, 22
Gene pool, 18
Genetic link, 19
Genetics, 18, 22
Grooming Anxieties, 45
Growling, 56
Hairless, 34
Hearing, 10
Heat, 20
Hissing, 56
Houseplants, 46
Hunting skills, 7, 15
Illness, 53
Instinct, 7, 15
Japanese Bobtail, 28
Karsh, Dr. Eileen, 12, 13
Kittenhood, 6, 8

Korat, 28
Life expectancy, 8
Litterbox, 48
Maine Coon Cat, 28
Manx, 30
Marking, 48
Maturation, 8
McCune, Dr. Sandra, 18
Mental stimuli, 7
Metabolic rate, 10
Milk, 43
Multiple handling, 13
Muscular function, 10
Neutering, 20, 22
Newborn, 5
Nocturnal, 62
Nourishment, 5
Nutritional needs, 43
Ocelot, 30
Ocicat, 30
Oriental Shorthair, 30
Overpopulation, 19
Parentage, 3
Pedigree, 19
Persian, 28, 31
Personality, 3, 7, 22, 23
Playing, 16, 53
Pregnancy, 7
Ragdoll, 31
Russian Blue, 33
Scandinavia, 30
Scent glands, 44
Scottish Fold, 33
Scratching post, 44, 57, 59
Sexual maturity, 7, 8
Shelters, 7
Siamese, 26, 30, 31, 33
Sight, 10
Socialization, 3, 12, 13, 15
Somali, 34
Spaying, 7, 22
Sphynx, 34
Spraying, 19
Spraying, 8, 48
Temple University, 12, 13
Territorial marking, 20
Tom cat, 7
Toys, 7
Training, 7
Turkish Angora, 34
Turkish Van, 34
Urination, 47
Veterinarian, 54
Waltham Centre for Pet Nutrition, 18, 19
Water, 43
Weaning, 6